The basic Sold

Learn how to solder
electronics effectively

Alex V. Punch

Table of Contents

CHAPTER ONE

THE MOST EFFECTIVE METHOD TO SOLDER: A BEGINNER'S GUIDE

Figuring out how to bind w/legitimate fastening procedures is a central expertise each creator ought to dominate. In this instructional exercise, we frame the fundamentals of fastening irons, binding stations, sorts of patch, desoldering and security tips. Whether you're fabricating a robot or working with Arduino, knowing how to patch will prove to be useful.

What Is Soldering?

If you somehow managed to dismantle any electronic gadget that contains a circuit board, you'll see the parts are connected utilizing binding methods. Binding is the most common way of joining at least two electronic parts together by dissolving weld around the association. Patch is a metal composite and when it cools it makes areas of strength for a connection between the parts. Despite the fact that fastening can make a long-lasting association, it can likewise be switched utilizing a desoldering instrument as portrayed underneath.

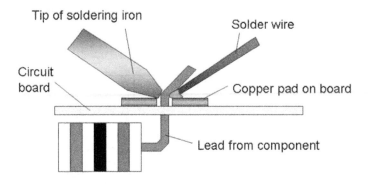

Tip of soldering iron

Solder wire

Circuit board

Copper pad on board

Lead from component

Fastening Tools

The beneficial thing about figuring out how to patch is the way that you needn't bother with a great deal to begin. Underneath we'll frame the essential instruments and materials you will require for the majority of your welding projects.

Patching Iron

A patching iron is a hand device that plugs into a standard 120v AC power source and warms up to soften weld around electrical associations. This is perhaps of the main device utilized in patching and it can arrive in a couple of varieties, for example, pen or firearm structure. For novices, it's suggested that you utilize the pen style fastening iron in the 15W to 30W territory. Most fastening irons have exchangeable tips that can be utilized for various patching applications. Be exceptionally careful while utilizing any sort of welding iron since it can warm up to 896' F which is very hot.

Patching Station

A patching station is a further developed form of the fundamental independent welding pen. On the off chance that you will be doing a ton of fastening, these are perfect to have as they offer greater adaptability and control. The primary advantage of a fastening station is the capacity to exactly change the temperature of the binding iron which is perfect for a scope of undertakings. These stations can likewise make a more secure work area as some incorporate high level temperature sensors, ready settings and even secret key insurance for wellbeing.

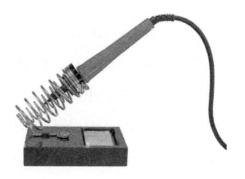

Patching Iron Tips

Toward the finish of most patching irons is an exchangeable part known as a binding tip. There are numerous varieties of this tip and they arrive in a wide assortment of shapes and sizes. Each tip is utilized for a particular reason and offers an unmistakable benefit over another. The most widely recognized tips you will use in gadgets projects are the conelike tip and the etch tip.

Tapered Tip - Used in accuracy hardware binding in view of the fine tip. As a result of its sharp end, conveying intensity to more modest regions without influencing its surroundings is capable.

Etch Tip - This tip is appropriate to patching wires or other bigger parts due to its wide level tip.

Brass or Conventional Sponge

Using a sponge can facilitate to stay the hand tool tip clean by removing the oxidisation that forms. Tips with oxidisation can tend to show black and not settle for solder because it did once it had been new. you may use a traditional wet sponge however this tends to shorten the time period of the tip because of enlargement and contraction. Also, a wet sponge can drop the temperature of the tip briefly once wiped. a far better different is to use a brass sponge as shown on the left.

Soldering Iron Stand

A hand tool stand terribly|is extremely|is incredibly} basic however very helpful and handy to own. This stand helps stop the new iron tip from returning involved with combustible materials or inflicting injury to your hand. Most bonding stations go with this in-built and additionally embrace a sponge or brass sponge for cleansing the tip.

Weld

Weld may be a metal compound material that's dissolved to create a particularly sturdy association between electrical components. It comes in each endlessly lead free varieties with breadths of .032″ and .062″ being the foremost well-known. within the bind center may be a material referred to as motion that works on contact and its mechanical strength. For gadgets binding, the foremost normally used sort is sans lead rosin center patch. this type of bind is mostly comprised of a Tin/Copper compound. you'll likewise utilize leaded 60/40 (60% tin, four-hundredth lead) rosin center weld but it's arising to be less acknowledge thanks to eudaemonia considerations. forward you actually do utilize lead bind, make sure you have applicable ventilation which you close up once use. While getting weld, strive to not utilize corrosive center bind as this can damage your circuits and components. Corrosive center weld is sold-out reception improvement stores and is usually used for plumbing and metal operating. As documented before, weld arrives in a minimum of a handful measurements. The thicker activity patch (.062″) is nice for fastening larger joints all the sooner nevertheless it will create fastening lesser joints hard. Therefore, it's typically very good to own the 2 sizes handy for your varied undertakings.

Assistance (Third Hand)

Some help may be a appliance that has a minimum of a pair of crocodilian reptile cuts and once during a whereas an amplifying glass/light connected. This clasps can assist you by holding the items you're trying to patch whereas you utilize the fastening iron and bind. AN exceptionally confirming device to own in your makerspace.

Patching Safety

Since it's currently therefore obvious what instruments and materials square measure needed, now's the correct time to momentarily examine approaches to remaining protected while binding.

Patching irons can arrive at temperatures of 800' F so it's vital to know where your iron is consistently. We generally suggest you utilize a patching iron stand to assist with forestalling incidental consumes or harm.

Ensure you are binding in a very much ventilated region. At the point when bind is warmed, there are exhaust delivered that are unsafe to your eyes and lungs. It's prescribed to utilize a smoke extractor which is a fan with a charcoal channel that ingests the destructive weld smoke you can visit destinations like Integrated Air Systems for air filtration frameworks. It's dependably really smart to wear defensive eye wear in the event of unplanned sprinkles of hot weld. In conclusion, make a point to clean up while done patching particularly on the off chance that utilizing lead bind.

Tinning The Tip

Before you can begin binding, you want to prepare your patching iron by tinning the tip with weld. This cycle will assist with further developing the intensity move from the iron to the thing you're welding. Tinning will likewise assist with safeguarding the tip and diminish wear.

Stage 1: Begin by ensuring the tip is connected to the iron and screwed firmly set up.

Stage 2: Turn on your welding iron and let it heat up. In the event that you have a binding station with a movable temp control, set it to 400′ C/752′ F.

Stage 3: Wipe the tip of the welding iron on a clammy wet wipe to clean it. Stand by a couple of moments to allow the tip to warm up again prior to continuing to stage 4.

Stage 4: Hold the welding iron in one hand and bind in the other. Contact the patch to the tip of the iron and ensure the bind streams equitably around the tip.

You ought to tin the tip of your iron when each welding meeting to broaden its life. At last, every tip will break down and will require supplanting when it turns out to be unpleasant or pitted.

CHAPTER TWO

STEP BY STEP INSTRUCTIONS TO SOLDER

To more readily clarify how for patch, we will show it with a certifiable application. In this model, we will patch a LED to a circuit board.

Stage 1: Mount The Component - Begin by embedding the leads of the LED into the openings of the circuit board. Flip the board over and twist the leads outward at a 45' point. This will assist the part make a superior association with the copper cushion and keep it from dropping out while welding.

Stage 2: Heat The Joint - Turn your welding iron on and assuming it has a flexible intensity control, set it to 400'C. Right now, contact the tip of the iron to the copper cushion and the resistor lead simultaneously. You want to hold the welding iron set up for 3-4 seconds to warm the cushion and the lead.

Stage 3: Apply Solder To Joint - Continue holding the welding iron on the copper cushion and the lead and contact your patch to the joint.

Significant - Don't contact the weld straightforwardly to the tip of the iron. You maintain that the joint should be adequately hot to dissolve the patch when it's contacted. In the event that the joint is excessively cool, it will shape a terrible association.

Stage 4: Snip The Leads - Remove the patching iron and let the weld cool down normally. Try not to blow on the patch as this will cause a terrible joint. When cool, you can cut the additional wire from leads. A legitimate patch joint is smooth, sparkling and seems to be a fountain of liquid magma or cone shape. You believe that barely enough weld should cover the whole joint yet not to an extreme so it turns into a ball or spills to a close by lead or joint.

CHAPTER THREE

HOW TO SOLDER WIRES

Now it's time to show you how to solder wires together. For this process, it's recommended to use helping hands or other type of clamp device. Begin by removing the insulation from the ends of both wires you are soldering together. If the wire is stranded, twist the strands together with your fingers.

1. Ensure your binding iron is completely warmed and contact the tip to the furthest

limit of one of the wires. Hold it on the wire for 3-4 seconds.

2. Keep the iron set up and contact the patch to the wire until it's completely covered. Rehash this cycle on the other wire.

3. Hold the two tinned wires on top of one another and contact the binding iron to the two wires. This interaction ought to liquefy the bind and coat the two wires equitably.

4. Eliminate the fastening iron and stand by a couple of

moments to allow the welded association with cool and solidify. Use heat psychologist to cover the association.

CHAPTER FOUR

DESOLDERING

The beneficial thing about utilizing patch is the way that it tends to be taken out effectively in a strategy known as desoldering. This proves to be useful on the off chance that you want to eliminate a part or make a revision to your electronic circuit. To desolder a joint, you will require patch wick which is otherwise called desoldering twist.

Stage 1 - Place a piece of the desoldering twist on top of the joint/weld you need eliminated.

Stage 2 - Heat your welding iron and contact the tip to the highest point of the mesh. This will warm the patch underneath which will then be consumed into the desoldering mesh. You can now eliminate the interlace to see the weld has been extricated and taken out. Be cautious contacting the twist when you are warming it since it will get hot.

Discretionary - If you have a ton of patch you need eliminated, you might need to utilize a gadget called a bind sucker. This is a handheld mechanical vacuum that sucks up hot weld with a press of a button. To utilize, press the unclogger down toward the finish of the patch sucker. Heat the joint with your welding iron and spot the tip of the patch sucker over the hot bind. Press the delivery button to suck up the fluid weld. To exhaust the weld sucker, push down on the unclogger.

Printed in Great Britain
by Amazon

37950831R00020